Greetings, bombers!

MARVELS 38
BOMBERS

BOMBERS

GRACE BELTOWSKI

CREATIVE EDUCATION | CREATIVE PAPERBACKS

READY FOR TAKEOFF!
Memphis Belle

table of contents

Greetings, Bombers! 1
Big Airplanes 7
Wings and Tail 8
Target Found 10
Flying a Bomber 12
Safe Flight. 14
What Do Bombers Do? 16
Farewell, Bombers! 18
Picture a Bomber 20
Words to Know 22
Read More 22
Websites 22
Index. 24

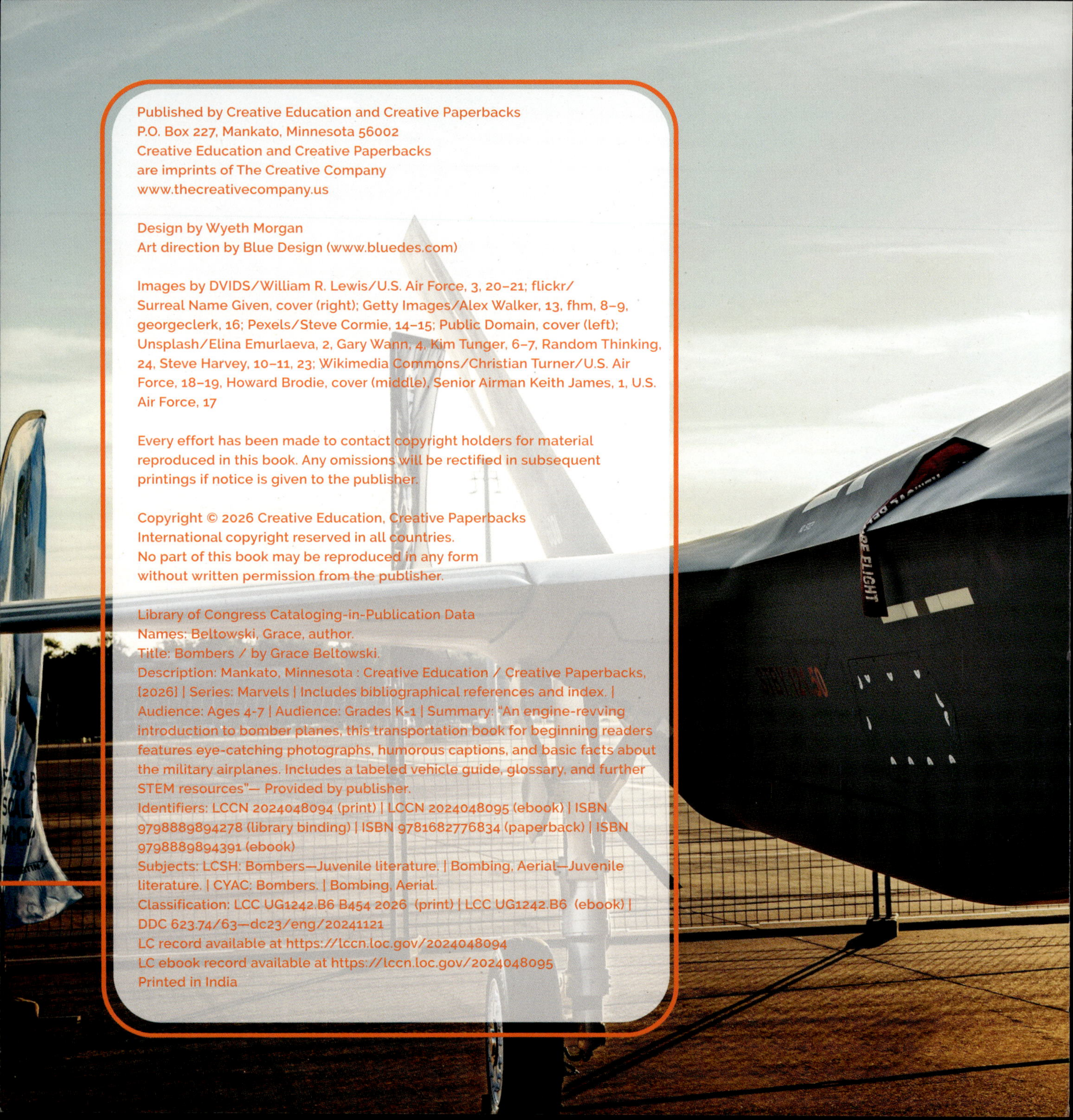

Published by Creative Education and Creative Paperbacks
P.O. Box 227, Mankato, Minnesota 56002
Creative Education and Creative Paperbacks
are imprints of The Creative Company
www.thecreativecompany.us

Design by Wyeth Morgan
Art direction by Blue Design (www.bluedes.com)

Images by DVIDS/William R. Lewis/U.S. Air Force, 3, 20–21; flickr/ Surreal Name Given, cover (right); Getty Images/Alex Walker, 13, fhm, 8–9, georgeclerk, 16; Pexels/Steve Cormie, 14–15; Public Domain, cover (left); Unsplash/Elina Emurlaeva, 2, Gary Wann, 4, Kim Tunger, 6–7, Random Thinking, 24, Steve Harvey, 10–11, 23; Wikimedia Commons/Christian Turner/U.S. Air Force, 18–19, Howard Brodie, cover (middle), Senior Airman Keith James, 1, U.S. Air Force, 17

Library of Congress Cataloging-in-Publication Data
Names: Beltowski, Grace, author.
Title: Bombers / by Grace Beltowski.
Description: Mankato, Minnesota : Creative Education / Creative Paperbacks, [2026] | Series: Marvels | Includes bibliographical references and index. | Audience: Ages 4-7 | Audience: Grades K-1 | Summary: "An engine-revving introduction to bomber planes, this transportation book for beginning readers features eye-catching photographs, humorous captions, and basic facts about the military airplanes. Includes a labeled vehicle guide, glossary, and further STEM resources"— Provided by publisher.
Identifiers: LCCN 2024048094 (print) | LCCN 2024048095 (ebook) | ISBN 9798889894278 (library binding) | ISBN 9781682776834 (paperback) | ISBN 9798889894391 (ebook)
Subjects: LCSH: Bombers—Juvenile literature. | Bombing, Aerial—Juvenile literature. | CYAC: Bombers. | Bombing, Aerial.
Classification: LCC UG1242.B6 B454 2026 (print) | LCC UG1242.B6 (ebook) | DDC 623.74/63—dc23/eng/20241121
LC record available at https://lccn.loc.gov/2024048094
LC ebook record available at https://lccn.loc.gov/2024048095
Printed in India

Bombers are big military airplanes. They drop bombs on enemy **targets**.

I AM A B-52 STRATOFORTRESS!

A bomber plane has two wings. The back of the plane is called the tail.

A bomber uses a computer to find targets. The computer shows where to drop the bombs.

SOME BOMBERS CAN DIVE IN THE AIR.

A **pilot** flies a bomber plane. The pilot sits in the **cockpit**.

I AM IMPORTANT!
SO AM I!

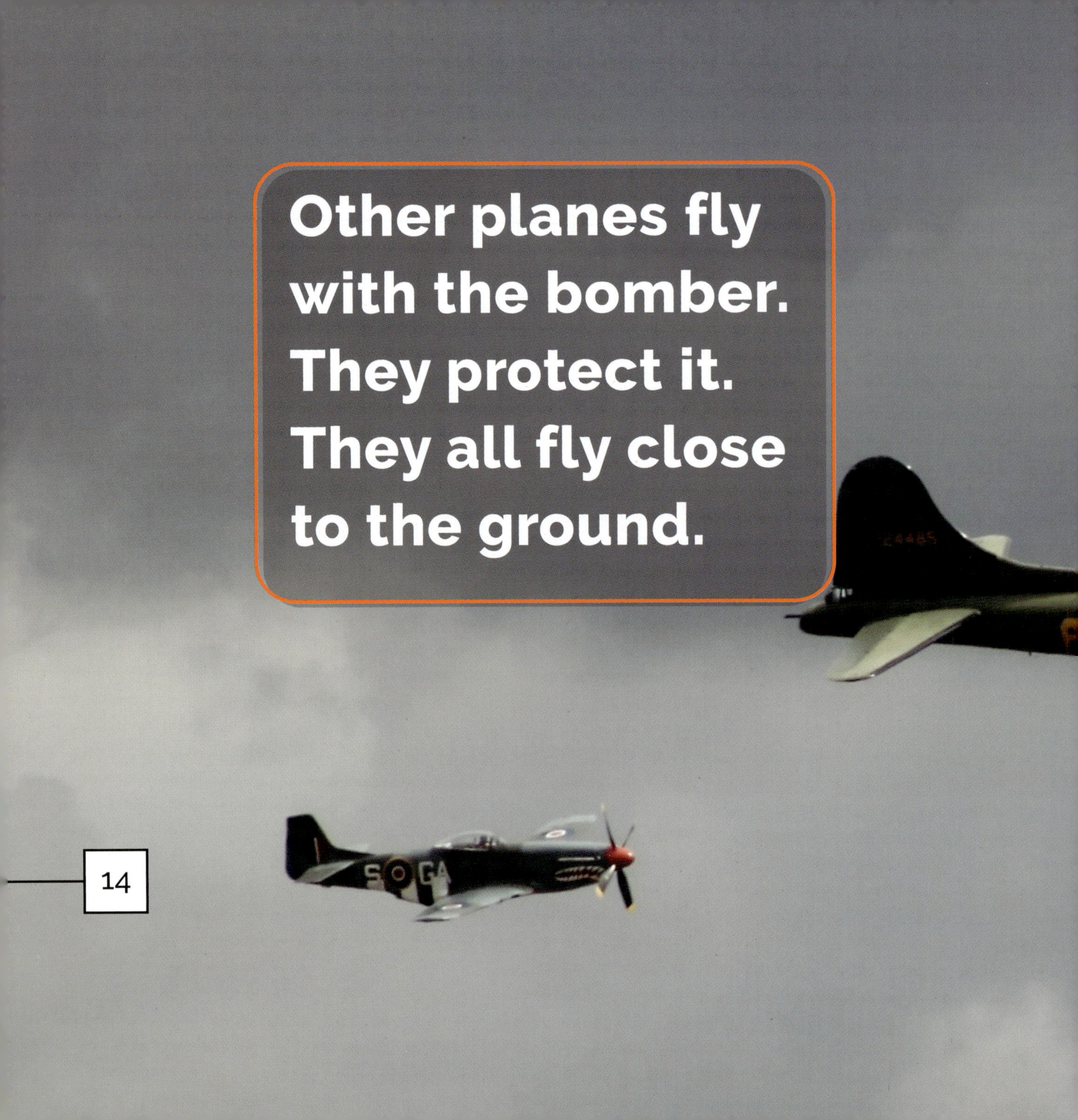

Other planes fly with the bomber. They protect it. They all fly close to the ground.

BOMBER PLANES ALSO CARRY GUNS.

A bomber plane finds its target. It opens a door to drop the bombs. It flies away quickly.

S. AIR FORCE

N842BA
21085

Farewell, bombers!

[Picture a Bomber]

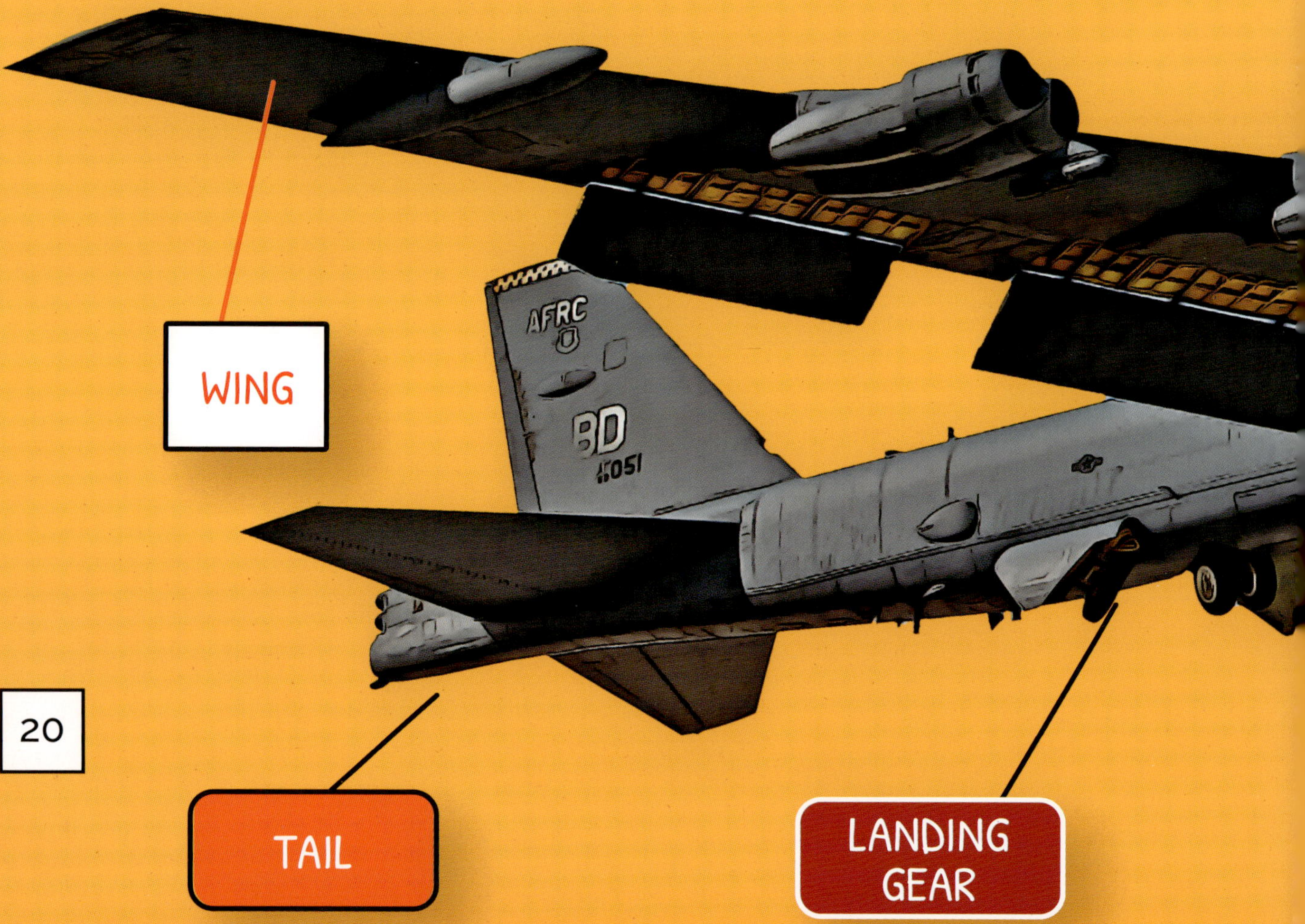

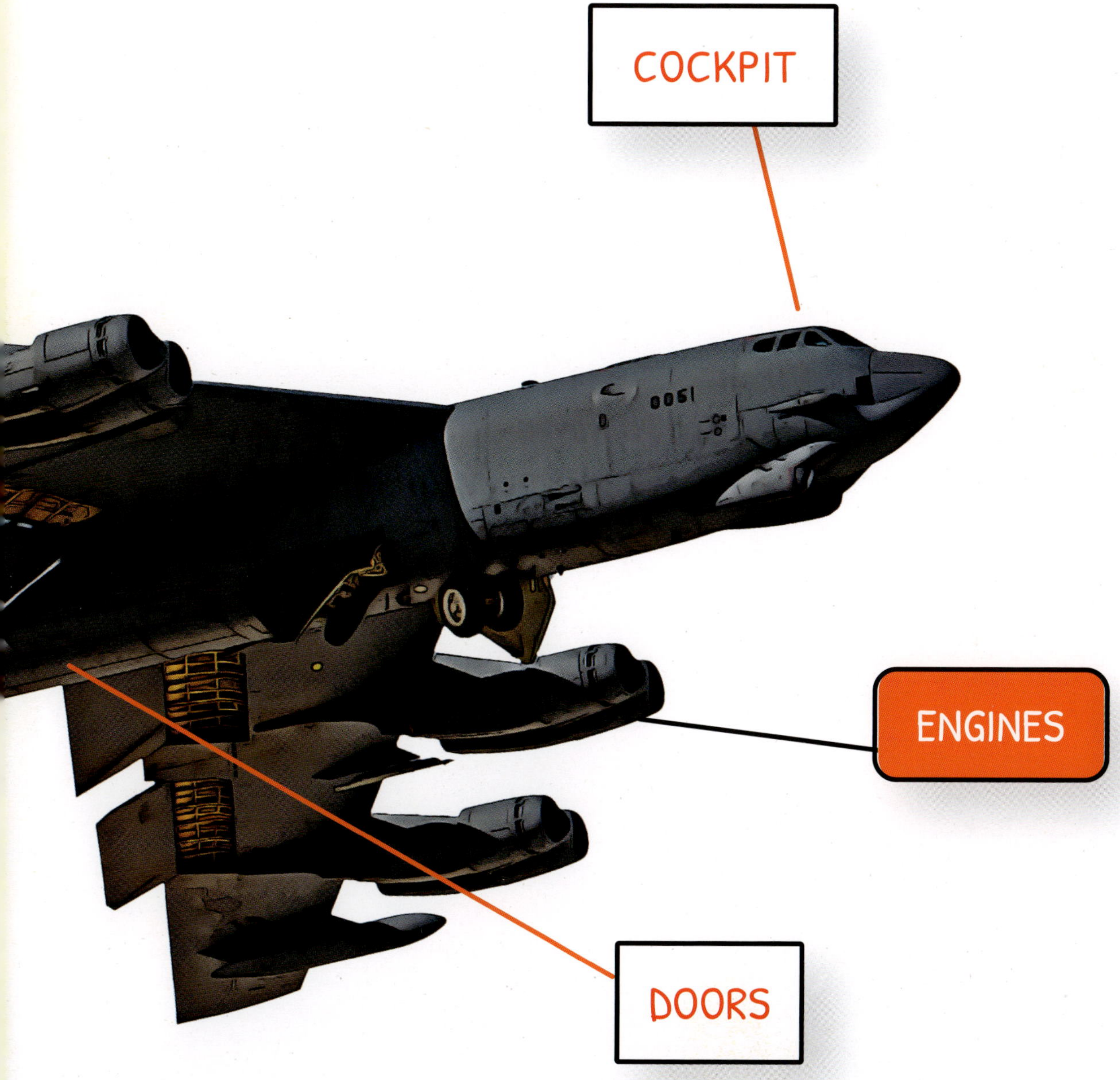
COCKPIT
ENGINES
DOORS

WORDS TO KNOW

cockpit: the place where an airplane pilot sits

pilot: the person in charge of flying a bomber plane

target: a plane, building, or another object that gets shot at

READ MORE

Bassier, Emma. *Military Aircraft*. Minneapolis: DiscoverRoo, 2020.

Franco, Michou. *I'm Going to Be a Pilot!* Buffalo, N.Y.: Gareth Stevens Publishing, 2024.

WEBSITES

Britannica Kids: Air Force
https://kids.britannica.com/kids/article/air-force/352717

Learn about the air force and how it uses bombers and other planes.

Kiddle: Bomber Facts for Kids
https://kids.kiddle.co/Bomber

Read about the history of bombers and look at photos of different military airplanes.

INDEX

bombs, 7, 10, 16

cockpits, 12, 21

computers, 10

doors, 16, 21

flying, 12, 14, 16

guns, 15

pilots, 12

tails, 9, 20

targets, 7, 10, 16

wings, 9, 20

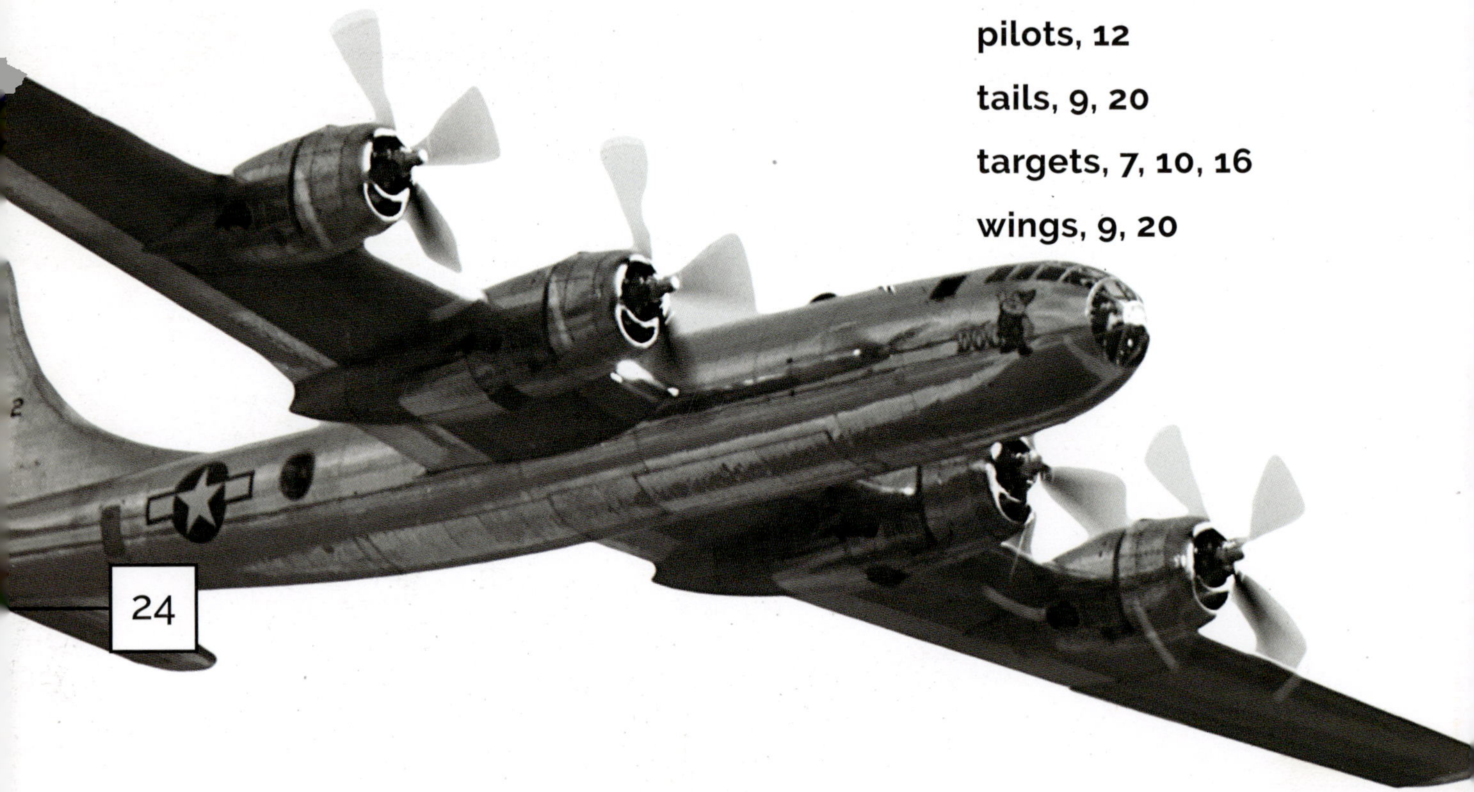